Name:	
Address:	
Phone:	Email:
VOLUME:	OF:

Astrophotography Journal
Astronomical Imaging
Field Log Edition

Date Start: / /	Location:	Moon Phase _____	
End: / /	GPS:	Brightness: _____ %	
Time Start: _____	Bortle Class: _____	Weather Conditions	
End: _____	Brightness: _____ %	🌡_____ 🚩_____ 💧_____	

Object Name:	Constellation:
Catalog Name:	Magnitude Size:
Scope:	Tracker:

Star Alignment: 1)   2)   3)

Camera:	Lens:	Filter/s:	Bahtinov Mask ☐
Exposure:	ISO:	Focal Length:	Aperture:

Lights #_____   Darks #_____   BIAS #_____   Flats #_____ | File Location

Notes:
_____
_____
_____
_____
_____
_____

Date	Start: / /	Location:	Moon Phase ___
	End: / /	GPS:	Brightness: ___ %
Time	Start: _____	Bortle Class: _____	Weather Conditions
	End: _____	Brightness: ___ %	🌡️_____ 🚩_____ 💧_____

Object Name:	Constellation:
Catalog Name:	Magnitude Size:
Scope:	Tracker:

Star Alignment: 1)   2)   3)

Camera:	Lens:	Filter/s:	Bahtinov Mask ☐
Exposure:	ISO:	Focal Length:	Aperture:

Lights # _____   Darks # _____   BIAS # _____   Flats # _____ | File Location

Notes:
_____
_____
_____
_____
_____
_____
_____

Date	Start: ___ / ___ / ___	Location:	Moon Phase _____
	End: ___ / ___ / ___	GPS:	Brightness: _____ %
Time	Start: _____	Bortle Class: _____	**Weather Conditions**
	End: _____	Brightness: _____ %	🌡 _____ 🚩 _____ 💧 _____

Object Name:	Constellation:
Catalog Name:	Magnitude Size:
Scope:	Tracker:

Star Alignment: 1)   2)   3)

Camera:	Lens:	Filter/s:	Bahtinov Mask ☐
Exposure:	ISO:	Focal Length:	Aperture:

Lights #_____   Darks #_____   BIAS #_____   Flats #_____   File Location

Notes:
_____
_____
_____
_____
_____
_____

_____
_____
_____
_____
_____
_____
_____
_____
_____

_____
_____
_____
_____
_____
_____
_____
_____
_____

Date Start: ___/___/___ End: ___/___/___	Location: GPS:	Moon Phase _____ Brightness: _____ %
Time Start: _____ End: _____	Bortle Class: _____ Brightness: _____ %	Weather Conditions 🌡️_____ 🚩_____ 💧_____

Object Name:	Constellation:
Catalog Name:	Magnitude Size:
Scope:	Tracker:

Star Alignment: 1)   2)   3)

Camera:	Lens:	Filter/s:	Bahtinov Mask ☐
Exposure:	ISO:	Focal Length:	Aperture:

Lights #_____   Darks #_____   BIAS #_____   Flats #_____ | File Location

Notes:
_____
_____
_____
_____
_____
_____

_____
_____
_____
_____
_____
_____
_____
_____
_____

_____
_____
_____
_____
_____
_____
_____
_____
_____
_____

Date Start: ___/___/___ End: ___/___/___	Location: GPS:	Moon Phase _____ Brightness: _____ %	
Time Start: _____ End: _____	Bortle Class: _____ Brightness: _____ %	**Weather Conditions** 🌡 _____ 🚩 _____ 💧 _____	

Object Name:	Constellation:
Catalog Name:	Magnitude Size:
Scope:	Tracker:

Star Alignment: 1)   2)   3)

Camera:	Lens:	Filter/s:	Bahtinov Mask ☐
Exposure:	ISO:	Focal Length:	Aperture:

Lights # _____   Darks # _____   BIAS # _____   Flats # _____   File Location

Notes:
_____
_____
_____
_____
_____

Date Start: ___/___/___	Location:		Moon Phase _____	
End: ___/___/___	GPS:		Brightness: _____ %	
Time Start: _____	Bortle Class: _____		**Weather Conditions**	
End: _____	Brightness: _____%	🌡_____	🚩_____	💧_____

Object Name:	Constellation:
Catalog Name:	Magnitude Size:
Scope:	Tracker:

Star Alignment: 1)   2)   3)

Camera:	Lens:	Filter/s:	Bahtinov Mask ☐
Exposure:	ISO:	Focal Length:	Aperture:

Lights #_____   Darks #_____   BIAS #_____   Flats #_____   File Location

Notes:
_____
_____
_____
_____
_____
_____

_____
_____
_____
_____
_____
_____
_____
_____
_____

_____
_____
_____
_____
_____
_____
_____
_____
_____
_____
_____

Date	Start: ___/___/___	Location:	Moon Phase _____
	End: ___/___/___	GPS:	Brightness: _____ %
Time	Start: _____	Bortle Class: _____	Weather Conditions
	End: _____	Brightness: _____%	🌡_____ 🚩_____ 💧_____

Object Name:	Constellation:
Catalog Name:	Magnitude Size:
Scope:	Tracker:

Star Alignment: 1)   2)   3)

| Camera: | Lens: | Filter/s: | Bahtinov Mask ☐ |
| Exposure: | ISO: | Focal Length: | Aperture: |

Lights # _____   Darks # _____   BIAS # _____   Flats # _____ | File Location

Notes:
_____
_____
_____
_____
_____
_____
_____

	Start: ___/___/___	Location:	Moon Phase _____
Date	End: ___/___/___	GPS:	Brightness: _____ %
Time	Start: _____	Bortle Class: _____	Weather Conditions
	End: _____	Brightness: _____%	🌡_____ 🚩_____ 💧_____

Object Name:	Constellation:
Catalog Name:	Magnitude Size:
Scope:	Tracker:

Star Alignment: 1)   2)   3)

Camera:	Lens:	Filter/s:	Bahtinov Mask ☐
Exposure:	ISO:	Focal Length:	Aperture:

Lights #_____   Darks #_____   BIAS #_____   Flats #_____   File Location

Notes:
_____
_____
_____
_____
_____
_____
_____

_____
_____
_____
_____
_____
_____
_____
_____

_____
_____
_____
_____
_____
_____
_____
_____

Date	Start: ___/___/___	Location:	Moon Phase _____
	End: ___/___/___	GPS:	Brightness: _____ %
Time	Start: _____	Bortle Class: _____	Weather Conditions
	End: _____	Brightness: _____ %	🌡_____ 🚩_____ 💧_____

Object Name:	Constellation:
Catalog Name:	Magnitude Size:
Scope:	Tracker:

Star Alignment: 1)   2)   3)

Camera:	Lens:	Filter/s:	Bahtinov Mask ☐
Exposure:	ISO:	Focal Length:	Aperture:

Lights #_____   Darks #_____   BIAS #_____   Flats #_____ | File Location

Notes: _____
_____
_____
_____
_____
_____
_____

_____
_____
_____
_____
_____
_____
_____
_____

_____
_____
_____
_____
_____
_____
_____
_____

	Start: / /	Location:	Moon Phase ____
Date	End: / /	GPS:	Brightness: ____ %
Time	Start: ____	Bortle Class: ____	Weather Conditions
	End: ____	Brightness: ____ %	🌡____ 🚩____ 💧____

Object Name:	Constellation:
Catalog Name:	Magnitude Size:
Scope:	Tracker:

Star Alignment: 1)   2)   3)

Camera:	Lens:	Filter/s:	Bahtinov Mask ☐
Exposure:	ISO:	Focal Length:	Aperture:

| Lights # ____ | Darks # ____ | BIAS # ____ | Flats # ____ | File Location |

Notes:
_____
_____
_____
_____
_____
_____

Date	Start: ___ / ___ / ___	Location:	Moon Phase _____
	End: ___ / ___ / ___	GPS:	Brightness: _____ %
Time	Start: _____	Bortle Class: _____	**Weather Conditions**
	End: _____	Brightness: _____ %	🌡️_____ 🚩_____ 💧_____

Object Name:	Constellation:
Catalog Name:	Magnitude Size:
Scope:	Tracker:

Star Alignment: 1)   2)   3)

| Camera: | Lens: | Filter/s: | Bahtinov Mask ☐ |
| Exposure: | ISO: | Focal Length: | Aperture: |

Lights # _____   Darks # _____   BIAS # _____   Flats # _____ | File Location

Notes:
_____
_____
_____
_____
_____
_____

_____
_____
_____
_____
_____
_____
_____
_____
_____

_____
_____
_____
_____
_____
_____
_____
_____

Date Start: ___ / ___ / ___	Location:	Moon Phase _____	
Date End: ___ / ___ / ___	GPS:	Brightness: _____ %	
Time Start: _____	Bortle Class: _____	Weather Conditions	
Time End: _____	Brightness: _____ %	🌡_____ ⚑_____ 💧_____	

Object Name:	Constellation:
Catalog Name:	Magnitude Size:
Scope:	Tracker:

Star Alignment: 1)   2)   3)

Camera:	Lens:	Filter/s:	Bahtinov Mask ☐
Exposure:	ISO:	Focal Length:	Aperture:

Lights # _____   Darks # _____   BIAS # _____   Flats # _____   File Location

Notes:
_____
_____
_____
_____
_____
_____

Date	Start: ___/___/___	Location:	Moon Phase _____
	End: ___/___/___	GPS:	Brightness: _____%
Time	Start: _____	Bortle Class: _____	Weather Conditions
	End: _____	Brightness: _____%	🌡_____ 🚩_____ 💧_____

Object Name:		Constellation:	
Catalog Name:		Magnitude Size:	
Scope:		Tracker:	

Star Alignment: 1) _____ 2) _____ 3) _____

Camera:	Lens:	Filter/s:	Bahtinov Mask ☐
Exposure:	ISO:	Focal Length:	Aperture:

Lights # _____   Darks # _____   BIAS # _____   Flats # _____   File Location

Notes:
_____
_____
_____
_____
_____
_____

_____
_____
_____
_____
_____
_____
_____
_____
_____

_____
_____
_____
_____
_____
_____
_____
_____
_____

Date	Start: ___ / ___ / ___	Location:	Moon Phase _____
	End: ___ / ___ / ___	GPS:	Brightness: _____ %
Time	Start: _____	Bortle Class: _____	Weather Conditions
	End: _____	Brightness: _____ %	🌡️ _____ 🚩 _____ 💧 _____

Object Name:	Constellation:
Catalog Name:	Magnitude Size:
Scope:	Tracker:

Star Alignment: 1) _____ 2) _____ 3) _____

Camera:	Lens:	Filter/s:	Bahtinov Mask ☐
Exposure:	ISO:	Focal Length:	Aperture:

Lights # _____ Darks # _____ BIAS # _____ Flats # _____ | File Location

Notes:
_____
_____
_____
_____
_____
_____

_____
_____
_____
_____
_____
_____
_____
_____

_____
_____
_____
_____
_____
_____
_____
_____

Date	Start: / /	Location:	Moon Phase _____
	End: / /	GPS:	Brightness: _____ %
Time	Start: _____	Bortle Class: _____	Weather Conditions
	End: _____	Brightness: _____ %	🌡_____ 🚩_____ 💧_____

Object Name:	Constellation:
Catalog Name:	Magnitude Size:
Scope:	Tracker:

Star Alignment: 1)   2)   3)

Camera:	Lens:	Filter/s:	Bahtinov Mask ☐
Exposure:	ISO:	Focal Length:	Aperture:

Lights #_____   Darks #_____   BIAS #_____   Flats #_____ | File Location

Notes:
_____
_____
_____
_____
_____
_____

Date	Start: / /	Location:	Moon Phase _____
	End: / /	GPS:	Brightness: _____ %
Time	Start: _____	Bortle Class: _____	Weather Conditions
	End: _____	Brightness: _____ %	🌡_____ 🚩_____ 💧_____

Object Name:	Constellation:
Catalog Name:	Magnitude Size:
Scope:	Tracker:

Star Alignment: 1)   2)   3)

Camera:	Lens:	Filter/s:	Bahtinov Mask ☐
Exposure:	ISO:	Focal Length:	Aperture:

Lights #_____   Darks #_____   BIAS #_____   Flats #_____   File Location

Notes:
_____
_____
_____
_____
_____
_____
_____

Date	Start: ___/___/___	Location:	Moon Phase _____
	End: ___/___/___	GPS:	Brightness: _____ %
Time	Start: _____	Bortle Class: _____	Weather Conditions
	End: _____	Brightness: _____ %	🌡___ 🚩___ 💧___

Object Name:	Constellation:
Catalog Name:	Magnitude Size:
Scope:	Tracker:

Star Alignment: 1)　　　　　2)　　　　　3)

Camera:	Lens:	Filter/s:	Bahtinov Mask ☐
Exposure:	ISO:	Focal Length:	Aperture:

Lights #_____　　Darks #_____　　BIAS #_____　　Flats #_____　| File Location

Notes:
_____
_____
_____
_____
_____
_____

_____
_____
_____
_____
_____
_____
_____
_____

_____
_____
_____
_____
_____
_____
_____

Date	Start: ___ / ___ / ___	Location:	Moon Phase ___
	End: ___ / ___ / ___	GPS:	Brightness: ___ %
Time	Start: _____	Bortle Class: _____	Weather Conditions
	End: _____	Brightness: _____ %	🌡___ 🚩___ 💧___

Object Name:	Constellation:
Catalog Name:	Magnitude Size:
Scope:	Tracker:

Star Alignment: 1)   2)   3)

Camera:	Lens:	Filter/s:	Bahtinov Mask ☐
Exposure:	ISO:	Focal Length:	Aperture:

Lights # _____   Darks # _____   BIAS # _____   Flats # _____ | File Location

Notes:
_____
_____
_____
_____
_____
_____
_____

_____
_____
_____
_____
_____
_____
_____
_____

_____
_____
_____
_____
_____
_____
_____
_____

Date	Start: ___ / ___ / ___	Location:	Moon Phase _____
	End: ___ / ___ / ___	GPS:	Brightness: _____ %
Time	Start: _____	Bortle Class: _____	Weather Conditions
	End: _____	Brightness: _____ %	🌡_____ 🚩_____ 💧_____

Object Name:	Constellation:
Catalog Name:	Magnitude Size:
Scope:	Tracker:

Star Alignment: 1)   2)   3)

| Camera: | Lens: | Filter/s: | Bahtinov Mask ☐ |
| Exposure: | ISO: | Focal Length: | Aperture: |

Lights #_____   Darks #_____   BIAS #_____   Flats #_____ | File Location

Notes:
_____
_____
_____
_____
_____
_____

Date	Start: ___/___/___	Location:	Moon Phase _____
	End: ___/___/___	GPS:	Brightness: _____ %
Time	Start: _____	Bortle Class: _____	Weather Conditions
	End: _____	Brightness: _____ %	🌡_____ 🚩_____ 💧_____

Object Name:	Constellation:
Catalog Name:	Magnitude Size:
Scope:	Tracker:

Star Alignment: 1)   2)   3)

Camera:	Lens:	Filter/s:	Bahtinov Mask ☐
Exposure:	ISO:	Focal Length:	Aperture:

Lights #_____   Darks #_____   BIAS #_____   Flats #_____   File Location

Notes:
_____
_____
_____
_____
_____

_____
_____
_____
_____
_____
_____
_____
_____
_____

_____
_____
_____
_____
_____
_____
_____
_____
_____

Date	Start: ___ / ___ / ___	Location:	Moon Phase _____
	End: ___ / ___ / ___	GPS:	Brightness: _____ %

Time	Start: _____	Bortle Class: _____	Weather Conditions		
	End: _____	Brightness: _____ %	🌡 _____	🚩 _____	💧 _____

Object Name:	Constellation:

Catalog Name:	Magnitude Size:

Scope:	Tracker:

Star Alignment: 1)   2)   3)

Camera:	Lens:	Filter/s:	Bahtinov Mask ☐
Exposure:	ISO:	Focal Length:	Aperture:

Lights # _____   Darks # _____   BIAS # _____   Flats # _____   File Location

Notes:
_____
_____
_____
_____
_____
_____

Date Start: / /	Location:	Moon Phase _____
Date End: / /	GPS:	Brightness: _____ %
Time Start: _____	Bortle Class: _____	Weather Conditions
Time End: _____	Brightness: _____ %	🌡_____ 🚩_____ 💧_____

Object Name:	Constellation:
Catalog Name:	Magnitude Size:
Scope:	Tracker:

Star Alignment: 1)   2)   3)

| Camera: | Lens: | Filter/s: | Bahtinov Mask ☐ |
| Exposure: | ISO: | Focal Length: | Aperture: |

Lights #_____   Darks #_____   BIAS #_____   Flats #_____ | File Location

Notes:
_____
_____
_____
_____
_____
_____

_____
_____
_____
_____
_____
_____
_____
_____

_____
_____
_____
_____
_____
_____
_____
_____

Date Start: ___/___/___ End: ___/___/___	Location: GPS:	Moon Phase _____ Brightness: _____ %	

| **Time** Start: _____ End: _____ | Bortle Class: _____ Brightness: _____% | Weather Conditions 🌡_____ 🚩_____ 💧_____ |

Object Name:	Constellation:
Catalog Name:	Magnitude Size:
Scope:	Tracker:

Star Alignment: 1)   2)   3)

Camera:	Lens:	Filter/s:	Bahtinov Mask ☐
Exposure:	ISO:	Focal Length:	Aperture:

Lights #_____   Darks #_____   BIAS #_____   Flats #_____   File Location

Notes:
_____
_____
_____
_____
_____
_____

Date	Start: / /	Location:	Moon Phase _____
	End: / /	GPS:	Brightness: _____ %
Time	Start: _____	Bortle Class: _____	Weather Conditions
	End: _____	Brightness: _____ %	🌡_____ 🚩_____ 💧_____

Object Name:	Constellation:
Catalog Name:	Magnitude Size:
Scope:	Tracker:

Star Alignment: 1)   2)   3)

| Camera: | Lens: | Filter/s: | Bahtinov Mask ☐ |
| Exposure: | ISO: | Focal Length: | Aperture: |

Lights # _____   Darks # _____   BIAS # _____   Flats # _____ | File Location

Notes:
_____
_____
_____
_____
_____
_____
_____

Date	Start: ___/___/___	Location:	Moon Phase _____
	End: ___/___/___	GPS:	Brightness: _____ %
Time	Start: _____	Bortle Class: _____	Weather Conditions
	End: _____	Brightness: _____ %	🌡 _____ 🎐 _____ 💧 _____

Object Name: _____   Constellation: _____

Catalog Name: _____   Magnitude Size: _____

Scope: _____   Tracker: _____

Star Alignment: 1) _____ 2) _____ 3) _____

| Camera: | Lens: | Filter/s: | Bahtinov Mask ☐ |
| Exposure: | ISO: | Focal Length: | Aperture: |

Lights #_____   Darks #_____   BIAS #_____   Flats #_____   File Location

Notes:
_____
_____
_____
_____
_____
_____

Date	Start: / /	Location:	Moon Phase
	End: / /	GPS:	Brightness: %
Time	Start:	Bortle Class:	Weather Conditions
	End:	Brightness: %	🌡 ___ 🚩 ___ 💧 ___

Object Name:	Constellation:
Catalog Name:	Magnitude Size:
Scope:	Tracker:

Star Alignment: 1)   2)   3)

| Camera: | Lens: | Filter/s: | Bahtinov Mask ☐ |
| Exposure: | ISO: | Focal Length: | Aperture: |

Lights #_____   Darks #_____   BIAS #_____   Flats #_____   File Location

Notes:
_____
_____
_____
_____
_____
_____

Date Start: ___ / ___ / ___	Location:	Moon Phase _____
Date End: ___ / ___ / ___	GPS:	Brightness: _____ %
Time Start: _____	Bortle Class: _____	**Weather Conditions**
Time End: _____	Brightness: _____%	🌡_____ 🚩_____ 💧_____

Object Name:	Constellation:
Catalog Name:	Magnitude Size:
Scope:	Tracker:

Star Alignment: 1)   2)   3)

Camera:	Lens:	Filter/s:	Bahtinov Mask ☐
Exposure:	ISO:	Focal Length:	Aperture:

Lights #_____   Darks #_____   BIAS #_____   Flats #_____   File Location

Notes:
_____
_____
_____
_____
_____
_____

Date Start: ___ / ___ / ___	Location:	Moon Phase ___
End: ___ / ___ / ___	GPS:	Brightness: ___ %
Time Start: ___	Bortle Class: ___	Weather Conditions
End: ___	Brightness: ___ %	🌡___ 🚩___ 💧___

Object Name:	Constellation:
Catalog Name:	Magnitude Size:
Scope:	Tracker:

Star Alignment: 1) _____ 2) _____ 3) _____

| Camera: | Lens: | Filter/s: | Bahtinov Mask ☐ |
| Exposure: | ISO: | Focal Length: | Aperture: |

Lights # _____   Darks # _____   BIAS # _____   Flats # _____ | File Location

Notes:
_____
_____
_____
_____
_____
_____

Date	Start: ___/___/___	Location:	Moon Phase _____
	End: ___/___/___	GPS:	Brightness: _____ %
Time	Start: _____	Bortle Class: _____	Weather Conditions
	End: _____	Brightness: _____ %	🌡_____ 🚩_____ ◌_____

Object Name:	Constellation:	
Catalog Name:	Magnitude Size:	
Scope:	Tracker:	

Star Alignment: 1) _____ 2) _____ 3) _____

Camera:	Lens:	Filter/s:	Bahtinov Mask ☐
Exposure:	ISO:	Focal Length:	Aperture:

Lights # _____   Darks # _____   BIAS # _____   Flats # _____   File Location

Notes:
_____
_____
_____
_____
_____
_____

Date	Start: ___/___/___	Location:	Moon Phase _____
	End: ___/___/___	GPS:	Brightness: _____ %
Time	Start: _____	Bortle Class: _____	Weather Conditions
	End: _____	Brightness: _____ %	🌡_____ 🚩_____ 💧_____

Object Name: _____ Constellation: _____

Catalog Name: _____ Magnitude Size: _____

Scope: _____ Tracker: _____

Star Alignment: 1) _____ 2) _____ 3) _____

| Camera: | Lens: | Filter/s: | Bahtinov Mask ☐ |
| Exposure: | ISO: | Focal Length: | Aperture: |

Lights #_____ Darks #_____ BIAS #_____ Flats #_____ File Location

Notes: _____
_____
_____
_____
_____
_____

	Start: ___/___/___	Location:	Moon Phase _____
Date	End: ___/___/___	GPS:	Brightness: _____ %
Time	Start: _____	Bortle Class: _____	**Weather Conditions**
	End: _____	Brightness: _____%	🌡___ 🚩_____ 💧_____

Object Name:	Constellation:
Catalog Name:	Magnitude Size:
Scope:	Tracker:

Star Alignment: 1)   2)   3)

Camera:	Lens:	Filter/s:	Bahtinov Mask ☐
Exposure:	ISO:	Focal Length:	Aperture:

Lights #_____   Darks #_____   BIAS #_____   Flats #_____ | File Location

Notes:
_____
_____
_____
_____
_____

Date	Start: / /	Location:	Moon Phase
	End: / /	GPS:	Brightness: %
Time	Start:	Bortle Class:	Weather Conditions
	End:	Brightness: %	🌡_____ 🚩_____ 💧_____

Object Name:	Constellation:
Catalog Name:	Magnitude Size:
Scope:	Tracker:

Star Alignment: 1)   2)   3)

Camera:	Lens:	Filter/s:	Bahtinov Mask ☐
Exposure:	ISO:	Focal Length:	Aperture:

Lights #_____   Darks #_____   BIAS #_____   Flats #_____   File Location

Notes:
_____
_____
_____
_____
_____
_____

Date	Start: / /	Location:	Moon Phase _____
	End: / /	GPS:	Brightness: _____ %
Time	Start: _____	Bortle Class: _____	Weather Conditions
	End: _____	Brightness: _____ %	🌡_____ 💨_____ 💧_____

Object Name:	Constellation:
Catalog Name:	Magnitude Size:
Scope:	Tracker:

Star Alignment: 1)   2)   3)

Camera:	Lens:	Filter/s:	Bahtinov Mask ☐
Exposure:	ISO:	Focal Length:	Aperture:

Lights #_____   Darks #_____   BIAS #_____   Flats #_____   File Location

Notes: _____
_____
_____
_____
_____

Date	Start: ___ / ___ / ___	Location:	Moon Phase _____
	End: ___ / ___ / ___	GPS:	Brightness: _____ %
Time	Start: _____	Bortle Class: _____	Weather Conditions
	End: _____	Brightness: _____ %	🌡_____ 🚩_____ 💧_____

Object Name:	Constellation:
Catalog Name:	Magnitude Size:
Scope:	Tracker:

Star Alignment: 1)   2)   3)

Camera:	Lens:	Filter/s:	Bahtinov Mask ☐
Exposure:	ISO:	Focal Length:	Aperture:

Lights #_____   Darks #_____   BIAS #_____   Flats #_____   File Location

Notes:
_____
_____
_____
_____
_____
_____

Date	Start: ___/___/___	Location:	Moon Phase _____
	End: ___/___/___	GPS:	Brightness: _____%

Time	Start: _____	Bortle Class: _____	Weather Conditions
	End: _____	Brightness: _____%	🌡_____ 🚩_____ 💧_____

Object Name:	Constellation:

Catalog Name:	Magnitude Size:

Scope:	Tracker:

Star Alignment: 1)   2)   3)

Camera:	Lens:	Filter/s:	Bahtinov Mask ☐
Exposure:	ISO:	Focal Length:	Aperture:

Lights #_____   Darks #_____   BIAS #_____   Flats #_____   File Location

Notes:
_____
_____
_____
_____
_____

_____
_____
_____
_____
_____
_____
_____
_____

_____
_____
_____
_____
_____
_____
_____
_____
_____

Date	Start: ___/___/___	Location:	Moon Phase _____
	End: ___/___/___	GPS:	Brightness: _____%
Time	Start: _____	Bortle Class: _____	Weather Conditions
	End: _____	Brightness: _____%	🌡_____ 🚩_____ 💧_____

Object Name:	Constellation:
Catalog Name:	Magnitude Size:
Scope:	Tracker:

Star Alignment: 1)   2)   3)

Camera:	Lens:	Filter/s:	Bahtinov Mask ☐
Exposure:	ISO:	Focal Length:	Aperture:

Lights #_____   Darks #_____   BIAS #_____   Flats #_____   File Location

Notes:
_____
_____
_____
_____
_____
_____

	Start: ___ / ___ / ___	Location:	Moon Phase _____
Date			
	End: ___ / ___ / ___	GPS:	Brightness: _____ %
Time	Start: _____	Bortle Class: _____	Weather Conditions
	End: _____	Brightness: _____ %	🌡___ 🚩___ 💧___

Object Name:	Constellation:
Catalog Name:	Magnitude Size:
Scope:	Tracker:

Star Alignment: 1) _____ 2) _____ 3) _____

| Camera: | Lens: | Filter/s: | Bahtinov Mask ☐ |
| Exposure: | ISO: | Focal Length: | Aperture: |

Lights #_____   Darks #_____   BIAS #_____   Flats #_____   File Location

Notes: _____
_____
_____
_____
_____
_____

Date Start: ___/___/___	Location:	Moon Phase _____
Date End: ___/___/___	GPS:	Brightness: _____ %
Time Start: _____	Bortle Class: _____	Weather Conditions
Time End: _____	Brightness: _____ %	🌡_____ 💨_____ 💧_____

Object Name:	Constellation:
Catalog Name:	Magnitude Size:
Scope:	Tracker:

Star Alignment: 1)   2)   3)

| Camera: | Lens: | Filter/s: | Bahtinov Mask ☐ |
| Exposure: | ISO: | Focal Length: | Aperture: |

Lights #_____   Darks #_____   BIAS #_____   Flats #_____   File Location

Notes:
_____
_____
_____
_____
_____
_____

_____
_____
_____
_____
_____
_____
_____
_____

_____
_____
_____
_____
_____
_____
_____
_____

| Date | Start: / / | Location: | Moon Phase _____ |
| | End: / / | GPS: | Brightness: _____ % |

| Time | Start: _____ | Bortle Class: _____ | Weather Conditions |
| | End: _____ | Brightness: _____ % | 🌡_____ 🚩_____ 💧_____ |

Object Name: Constellation:

Catalog Name: Magnitude Size:

Scope: Tracker:

Star Alignment: 1)   2)   3)

| Camera: | Lens: | Filter/s: | Bahtinov Mask ☐ |
| Exposure: | ISO: | Focal Length: | Aperture: |

Lights #_____   Darks #_____   BIAS #_____   Flats #_____   File Location

Notes:
_____
_____
_____
_____
_____
_____

_____
_____
_____
_____
_____
_____
_____
_____

_____
_____
_____
_____
_____
_____
_____
_____
_____

Date	Start: ___ / ___ / ___	Location:	Moon Phase _____
	End: ___ / ___ / ___	GPS:	Brightness: _____ %
Time	Start: _____	Bortle Class: _____	Weather Conditions
	End: _____	Brightness: _____ %	🌡____ 🚩____ 💧____

Object Name:	Constellation:
Catalog Name:	Magnitude Size:
Scope:	Tracker:

Star Alignment: 1)　　　　　　　2)　　　　　　　3)

Camera:	Lens:	Filter/s:	Bahtinov Mask ☐
Exposure:	ISO:	Focal Length:	Aperture:

Lights #_____　　Darks #_____　　BIAS #_____　　Flats #_____　　File Location

Notes:
_____
_____
_____
_____
_____
_____

_____
_____
_____
_____
_____
_____
_____
_____

_____
_____
_____
_____
_____
_____
_____
_____
_____

Date	Start: ___ / ___ / ___	Location:	Moon Phase _____
	End: ___ / ___ / ___	GPS:	Brightness: _____ %
Time	Start: _____	Bortle Class: _____	Weather Conditions
	End: _____	Brightness: _____ %	🌡_____ 🚩_____ 💧_____

Object Name:	Constellation:
Catalog Name:	Magnitude Size:
Scope:	Tracker:

Star Alignment: 1)　　　　　　　　2)　　　　　　　　3)

Camera:	Lens:	Filter/s:	Bahtinov Mask ☐
Exposure:	ISO:	Focal Length:	Aperture:

Lights #_____　　Darks #_____　　BIAS #_____　　Flats #_____　　File Location

Notes:
_____
_____
_____
_____
_____
_____

_____
_____
_____
_____
_____
_____
_____

_____
_____
_____
_____
_____
_____
_____
_____
_____

Date	Start: ___/___/___	Location:	Moon Phase _____
	End: ___/___/___	GPS:	Brightness: _____ %
Time	Start: _____	Bortle Class: _____	Weather Conditions
	End: _____	Brightness: _____%	🌡_____ 🚩_____ 💧_____

Object Name: **Constellation:**

Catalog Name: **Magnitude Size:**

Scope: **Tracker:**

Star Alignment: 1)   **2)**   **3)**

| Camera: | Lens: | Filter/s: | Bahtinov Mask ☐ |
| Exposure: | ISO: | Focal Length: | Aperture: |

Lights #_____ Darks #_____ BIAS #_____ Flats #_____ File Location

Notes:
_____
_____
_____
_____
_____
_____

Date	Start: ___/___/___	Location:	Moon Phase _____
	End: ___/___/___	GPS:	Brightness: _____ %
Time	Start: _____	Bortle Class: _____	Weather Conditions
	End: _____	Brightness: _____%	🌡_____ ⚑_____ 💧_____

Object Name:	Constellation:
Catalog Name:	Magnitude Size:
Scope:	Tracker:

Star Alignment: 1)   2)   3)

Camera:	Lens:	Filter/s:	Bahtinov Mask ☐
Exposure:	ISO:	Focal Length:	Aperture:
Lights # _____	Darks # _____	BIAS # _____	Flats # _____	File Location

Notes:
_____
_____
_____
_____
_____
_____

	Start: ___ / ___ / ___	Location:	Moon Phase ___
Date			
	End: ___ / ___ / ___	GPS:	Brightness: ___ %
Time	Start: _____	Bortle Class: _____	Weather Conditions
	End: _____	Brightness: _____ %	🌡___ 🚩___ ◌___

Object Name: **Constellation:**

Catalog Name: **Magnitude Size:**

Scope: **Tracker:**

Star Alignment: 1)   2)   3)

| Camera: | Lens: | Filter/s: | Bahtinov Mask ☐ |
| Exposure: | ISO: | Focal Length: | Aperture: |

Lights #_____   Darks #_____   BIAS #_____   Flats #_____   File Location

Notes:
_____
_____
_____
_____
_____
_____

Date	Start: ___/___/___	Location:	Moon Phase _____
	End: ___/___/___	GPS:	Brightness: _____ %
Time	Start: _____	Bortle Class: _____	Weather Conditions
	End: _____	Brightness: _____ %	🌡_____ 🚩_____ 💧_____

Object Name:	Constellation:
Catalog Name:	Magnitude Size:
Scope:	Tracker:

Star Alignment: 1)   2)   3)

Camera:	Lens:	Filter/s:	Bahtinov Mask ☐
Exposure:	ISO:	Focal Length:	Aperture:

Lights # _____   Darks # _____   BIAS # _____   Flats # _____   File Location

Notes:

Date	Start: ___/___/___	Location:	Moon Phase _____
	End: ___/___/___	GPS:	Brightness: _____ %

Time	Start: _____	Bortle Class: _____	Weather Conditions
	End: _____	Brightness: _____ %	🌡_____ 🚩_____ 💧_____

Object Name:	Constellation:
Catalog Name:	Magnitude Size:
Scope:	Tracker:

Star Alignment: 1)   2)   3)

Camera:	Lens:	Filter/s:	Bahtinov Mask ☐
Exposure:	ISO:	Focal Length:	Aperture:

Lights #_____   Darks #_____   BIAS #_____   Flats #_____   File Location

Notes:
_____
_____
_____
_____
_____
_____

Date	Start: ___/___/___	Location:	Moon Phase _____
	End: ___/___/___	GPS:	Brightness: _____ %
Time	Start: _____	Bortle Class: _____	Weather Conditions
	End: _____	Brightness: _____%	🌡_____ ⚐_____ ⬙_____

Object Name:	Constellation:
Catalog Name:	Magnitude Size:
Scope:	Tracker:

Star Alignment: 1) _____ 2) _____ 3) _____

Camera:	Lens:	Filter/s:	Bahtinov Mask ☐
Exposure:	ISO:	Focal Length:	Aperture:

Lights #_____   Darks #_____   BIAS #_____   Flats #_____ | File Location

Notes:
_____
_____
_____
_____
_____

Date	Start: / /	Location:	Moon Phase _____
	End: / /	GPS:	Brightness: _____ %
Time	Start: _____	Bortle Class: _____	Weather Conditions
	End: _____	Brightness: _____ %	🌡_____ 🚩_____ 💧_____

Object Name:	Constellation:
Catalog Name:	Magnitude Size:
Scope:	Tracker:

Star Alignment: 1)   2)   3)

| Camera: | Lens: | Filter/s: | Bahtinov Mask ☐ |
| Exposure: | ISO: | Focal Length: | Aperture: |

Lights #_____   Darks #_____   BIAS #_____   Flats #_____   File Location

Notes:
_____
_____
_____
_____
_____
_____

_____
_____
_____
_____
_____
_____
_____
_____

_____
_____
_____
_____
_____
_____
_____
_____
_____
_____

Date	Start: ___/___/___	Location:	Moon Phase _____
	End: ___/___/___	GPS:	Brightness: _____ %
Time	Start: _____	Bortle Class: _____	**Weather Conditions**
	End: _____	Brightness: _____ %	🌡_____ 🚩_____ 💧_____

Object Name:	Constellation:
Catalog Name:	Magnitude Size:
Scope:	Tracker:

Star Alignment: 1)   2)   3)

| Camera: | Lens: | Filter/s: | Bahtinov Mask ☐ |
| Exposure: | ISO: | Focal Length: | Aperture: |

Lights # _____   Darks # _____   BIAS # _____   Flats # _____   File Location

Notes:
_____
_____
_____
_____
_____

Date	Start: ___/___/___	Location:	Moon Phase _____
	End: ___/___/___	GPS:	Brightness: _____ %
Time	Start: _____	Bortle Class: _____	Weather Conditions
	End: _____	Brightness: _____ %	🌡_____ 💨_____ 💧_____

Object Name: _____   Constellation: _____

Catalog Name: _____   Magnitude Size: _____

Scope: _____   Tracker: _____

Star Alignment: 1) _____   2) _____   3) _____

| Camera: | Lens: | Filter/s: | Bahtinov Mask ☐ |
| Exposure: | ISO: | Focal Length: | Aperture: |

Lights #_____   Darks #_____   BIAS #_____   Flats #_____ | File Location

Notes:
_____
_____
_____
_____
_____
_____

	Start: / /	Location:	Moon Phase _____
Date	End: / /	GPS:	Brightness: _____ %
Time	Start: _____	Bortle Class: _____	Weather Conditions
	End: _____	Brightness: _____ %	🌡_____ 🚩_____ 💧_____

Object Name:	Constellation:
Catalog Name:	Magnitude Size:
Scope:	Tracker:

Star Alignment: 1)   2)   3)

| Camera: | Lens: | Filter/s: | Bahtinov Mask ☐ |
| Exposure: | ISO: | Focal Length: | Aperture: |

Lights #_____   Darks #_____   BIAS #_____   Flats #_____ | File Location

Notes:
_____
_____
_____
_____
_____
_____

Date Start: / /	Location:	Moon Phase _____
Date End: / /	GPS:	Brightness: _____ %
Time Start: _____	Bortle Class: _____	Weather Conditions
Time End: _____	Brightness: _____ %	🌡_____ 🚩_____ 💧_____

Object Name:	Constellation:
Catalog Name:	Magnitude Size:
Scope:	Tracker:

Star Alignment: 1)   2)   3)

Camera:	Lens:	Filter/s:	Bahtinov Mask ☐
Exposure:	ISO:	Focal Length:	Aperture:

Lights #_____   Darks #_____   BIAS #_____   Flats #_____ | File Location

Notes:
_____
_____
_____
_____
_____
_____

Date	Start: ___/___/___	Location:	Moon Phase _____
	End: ___/___/___	GPS:	Brightness: _____%
Time	Start: _____	Bortle Class: _____	Weather Conditions
	End: _____	Brightness: _____%	🌡_____ ⚑_____ 💧_____

Object Name: _____ Constellation: _____

Catalog Name: _____ Magnitude Size: _____

Scope: _____ Tracker: _____

Star Alignment: 1) _____ 2) _____ 3) _____

| Camera: | Lens: | Filter/s: | Bahtinov Mask ☐ |
| Exposure: | ISO: | Focal Length: | Aperture: |

Lights #_____ Darks #_____ BIAS #_____ Flats #_____ File Location

Notes:
_____
_____
_____
_____
_____

Date	Start: ___ / ___ / ___	Location:	Moon Phase _____
	End: ___ / ___ / ___	GPS:	Brightness: _____ %
Time	Start: _____	Bortle Class: _____	Weather Conditions
	End: _____	Brightness: _____ %	🌡_____ 🎏_____ 💧_____

Object Name:	Constellation:
Catalog Name:	Magnitude Size:
Scope:	Tracker:

Star Alignment: 1)   2)   3)

Camera:	Lens:	Filter/s:	Bahtinov Mask ☐
Exposure:	ISO:	Focal Length:	Aperture:

Lights #_____   Darks #_____   BIAS #_____   Flats #_____ | File Location

Notes:
_____
_____
_____
_____
_____
_____

_____
_____
_____
_____
_____
_____
_____
_____
_____
_____

_____
_____
_____
_____
_____
_____
_____
_____
_____
_____

Date	Start: ___/___/___	Location:	Moon Phase _____
	End: ___/___/___	GPS:	Brightness: _____ %
Time	Start: _____	Bortle Class: _____	Weather Conditions
	End: _____	Brightness: _____%	🌡_____ 🚩_____ 💧_____

Object Name: Constellation:

Catalog Name: Magnitude Size:

Scope: Tracker:

Star Alignment: 1)   2)   3)

Camera:	Lens:	Filter/s:	Bahtinov Mask ☐
Exposure:	ISO:	Focal Length:	Aperture:

Lights #_____   Darks #_____   BIAS #_____   Flats #_____   File Location

Notes:

_____

_____

_____

_____

_____

Date	Start: ___/___/___	Location:	Moon Phase _____
	End: ___/___/___	GPS:	Brightness: _____ %
Time	Start: _____	Bortle Class: _____	Weather Conditions
	End: _____	Brightness: _____ %	🌡_____ 🚩_____ 💧_____

Object Name:	Constellation:
Catalog Name:	Magnitude Size:
Scope:	Tracker:

Star Alignment: 1)   2)   3)

Camera:	Lens:	Filter/s:	Bahtinov Mask ☐
Exposure:	ISO:	Focal Length:	Aperture:

Lights #_____   Darks #_____   BIAS #_____   Flats #_____   File Location

Notes:
_____
_____
_____
_____
_____
_____

_____
_____
_____
_____
_____
_____
_____
_____

_____
_____
_____
_____
_____
_____
_____
_____

Date	Start: / /	Location:	Moon Phase _____
	End: / /	GPS:	Brightness: _____ %
Time	Start: _____	Bortle Class: _____	Weather Conditions
	End: _____	Brightness: _____ %	🌡_____ 🚩_____ 💧_____

Object Name:	Constellation:
Catalog Name:	Magnitude Size:
Scope:	Tracker:

Star Alignment: 1)   2)   3)

Camera:	Lens:	Filter/s:	Bahtinov Mask ☐
Exposure:	ISO:	Focal Length:	Aperture:

Lights #_____   Darks #_____   BIAS #_____   Flats #_____   File Location

Notes:
_____
_____
_____
_____
_____

Date	Start: ___/___/___	Location:	Moon Phase _____
	End: ___/___/___	GPS:	Brightness: _____%
Time	Start: _____	Bortle Class: _____	Weather Conditions
	End: _____	Brightness: _____%	🌡_____ 🚩_____ 💧_____

Object Name:	Constellation:
Catalog Name:	Magnitude Size:
Scope:	Tracker:

Star Alignment: 1)   2)   3)

Camera:	Lens:	Filter/s:	Bahtinov Mask ☐
Exposure:	ISO:	Focal Length:	Aperture:

Lights #_____   Darks #_____   BIAS #_____   Flats #_____ | File Location

Notes:
_____
_____
_____
_____
_____
_____

Date	Start: ___ / ___ / ___	Location:	Moon Phase _____
	End: ___ / ___ / ___	GPS:	Brightness: _____ %
Time	Start: _____	Bortle Class: _____	Weather Conditions
	End: _____	Brightness: _____ %	🌡 _____ 💨 _____ 💧 _____

Object Name:	Constellation:
Catalog Name:	Magnitude Size:
Scope:	Tracker:

Star Alignment: 1)   2)   3)

Camera:	Lens:	Filter/s:	Bahtinov Mask ☐
Exposure:	ISO:	Focal Length:	Aperture:

Lights #_____   Darks #_____   BIAS #_____   Flats #_____ | File Location

Notes:
_____
_____
_____
_____
_____
_____

Date	Start: ___ / ___ / ___	Location:	Moon Phase _____
	End: ___ / ___ / ___	GPS:	Brightness: _____ %
Time	Start: _____	Bortle Class: _____	Weather Conditions
	End: _____	Brightness: _____ %	🌡_____ 🚩_____ 💧_____

Object Name:	Constellation:
Catalog Name:	Magnitude Size:
Scope:	Tracker:

Star Alignment: 1)   2)   3)

Camera:	Lens:	Filter/s:	Bahtinov Mask ☐
Exposure:	ISO:	Focal Length:	Aperture:

Lights # _____   Darks # _____   BIAS # _____   Flats # _____ | File Location

Notes:
_____
_____
_____
_____
_____
_____

_____
_____
_____
_____
_____
_____
_____
_____

_____
_____
_____
_____
_____
_____
_____
_____
_____

Date	Start: ___/___/___	Location:	Moon Phase _____
	End: ___/___/___	GPS:	Brightness: _____ %
Time	Start: _____	Bortle Class: _____	Weather Conditions
	End: _____	Brightness: _____ %	🌡 _____ 🎏 _____ 💧 _____

Object Name:	Constellation:
Catalog Name:	Magnitude Size:
Scope:	Tracker:

Star Alignment: 1)   2)   3)

| Camera: | Lens: | Filter/s: | Bahtinov Mask ☐ |
| Exposure: | ISO: | Focal Length: | Aperture: |

Lights #_____   Darks #_____   BIAS #_____   Flats #_____   File Location

Notes:
_____
_____
_____
_____
_____
_____

_____
_____
_____
_____
_____
_____
_____

_____
_____
_____
_____
_____
_____
_____
_____
_____

Date Start: ___/___/___ End: ___/___/___	Location: GPS:	Moon Phase _____ Brightness: _____ %	

Time Start: _____ End: _____	Bortle Class: _____ Brightness: _____ %	Weather Conditions 🌡_____ 🚩_____ 💧_____

Object Name:	Constellation:
Catalog Name:	Magnitude Size:
Scope:	Tracker:

Star Alignment: 1)　　　　　　　　2)　　　　　　　　3)

Camera:	Lens:	Filter/s:	Bahtinov Mask ☐
Exposure:	ISO:	Focal Length:	Aperture:

Lights #_____　　Darks #_____　　BIAS #_____　　Flats #_____　　File Location

Notes:
_____
_____
_____
_____
_____
_____

Date	Start: ___/___/___	Location:	Moon Phase _____
	End: ___/___/___	GPS:	Brightness: _____ %
Time	Start: _____	Bortle Class: _____	Weather Conditions
	End: _____	Brightness: _____ %	🌡_____ 🚩_____ 💧_____

Object Name:	Constellation:
Catalog Name:	Magnitude Size:
Scope:	Tracker:

Star Alignment: 1)   2)   3)

Camera:	Lens:	Filter/s:	Bahtinov Mask ☐
Exposure:	ISO:	Focal Length:	Aperture:

Lights #_____   Darks #_____   BIAS #_____   Flats #_____   File Location

Notes:
_____
_____
_____
_____
_____
_____

Date	Start: / /	Location:	Moon Phase
	End: / /	GPS:	Brightness: %
Time	Start:	Bortle Class:	Weather Conditions
	End:	Brightness: %	🌡 🚩 💧

Object Name:	Constellation:
Catalog Name:	Magnitude Size:
Scope:	Tracker:

Star Alignment: 1)   2)   3)

Camera:	Lens:	Filter/s:	Bahtinov Mask ☐
Exposure:	ISO:	Focal Length:	Aperture:

Lights #_____   Darks #_____   BIAS #_____   Flats #_____   File Location

Notes:
_____
_____
_____
_____
_____
_____

Date	Start: ___/___/___	Location:	Moon Phase _____
	End: ___/___/___	GPS:	Brightness: _____ %
Time	Start: _____	Bortle Class: _____	Weather Conditions
	End: _____	Brightness: _____ %	🌡___ 🚩___ 💧___

Object Name:	Constellation:
Catalog Name:	Magnitude Size:
Scope:	Tracker:

Star Alignment: 1)　　　　　　　2)　　　　　　　3)

| Camera: | Lens: | Filter/s: | Bahtinov Mask ☐ |
| Exposure: | ISO: | Focal Length: | Aperture: |

Lights # _____　　Darks # _____　　BIAS # _____　　Flats # _____　| File Location

Notes:
_____
_____
_____
_____
_____

Date Start: ___/___/___	Location:	Moon Phase _____	
Date End: ___/___/___	GPS:	Brightness: _____ %	
Time Start: _____	Bortle Class: _____	**Weather Conditions**	
Time End: _____	Brightness: _____ %	🌡 _____ 🚩_____ 💧_____	

Object Name:	Constellation:
Catalog Name:	Magnitude Size:
Scope:	Tracker:

Star Alignment: 1)   2)   3)

Camera:	Lens:	Filter/s:	Bahtinov Mask ☐
Exposure:	ISO:	Focal Length:	Aperture:

Lights # _____   Darks # _____   BIAS # _____   Flats # _____ | File Location

Notes:
_____
_____
_____
_____
_____
_____
_____

Date	Start: ___ / ___ / ___	Location:	Moon Phase _____
	End: ___ / ___ / ___	GPS:	Brightness: _____ %
Time	Start: _____	Bortle Class: _____	Weather Conditions
	End: _____	Brightness: _____ %	🌡_____ 🚩_____ 💧_____

Object Name:	Constellation:
Catalog Name:	Magnitude Size:
Scope:	Tracker:

Star Alignment: 1)   2)   3)

Camera:	Lens:	Filter/s:	Bahtinov Mask ☐
Exposure:	ISO:	Focal Length:	Aperture:

Lights #_____   Darks #_____   BIAS #_____   Flats #_____   File Location

Notes:
_____
_____
_____
_____
_____
_____

Date	Start: ___ / ___ / ___	Location:	Moon Phase _____
	End: ___ / ___ / ___	GPS:	Brightness: _____ %

Time	Start: _____	Bortle Class: _____	Weather Conditions		
	End: _____	Brightness: _____ %	🌡 _____	🚩 _____	💧 _____

Object Name:	Constellation:
Catalog Name:	Magnitude Size:
Scope:	Tracker:

Star Alignment: 1) _____ 2) _____ 3) _____

Camera:	Lens:	Filter/s:	Bahtinov Mask ☐
Exposure:	ISO:	Focal Length:	Aperture:

Lights #_____   Darks #_____   BIAS #_____   Flats #_____   File Location

Notes:
_____
_____
_____
_____
_____

_____
_____
_____
_____
_____
_____
_____
_____

_____
_____
_____
_____
_____
_____
_____
_____
_____
_____

Date	Start: ___/___/___	Location:	Moon Phase _____
	End: ___/___/___	GPS:	Brightness: _____ %
Time	Start: _____	Bortle Class: _____	Weather Conditions
	End: _____	Brightness: _____ %	🌡_____ 🚩_____ 💧_____

Object Name:	Constellation:
Catalog Name:	Magnitude Size:
Scope:	Tracker:

Star Alignment: 1)   2)   3)

Camera:	Lens:	Filter/s:	Bahtinov Mask ☐
Exposure:	ISO:	Focal Length:	Aperture:

Lights # _____   Darks # _____   BIAS # _____   Flats # _____   File Location

Notes:
_____
_____
_____
_____
_____
_____

Date Start: ___/___/___ End: ___/___/___	Location: GPS:	Moon Phase _____ Brightness: _____ %	
Time Start: _____ End: _____	Bortle Class: _____ Brightness: _____ %	Weather Conditions 🌡_____ 🚩_____ 💧_____	

Object Name:	Constellation:
Catalog Name:	Magnitude Size:
Scope:	Tracker:

Star Alignment: 1)   2)   3)

Camera:	Lens:	Filter/s:	Bahtinov Mask ☐
Exposure:	ISO:	Focal Length:	Aperture:

Lights #_____   Darks #_____   BIAS #_____   Flats #_____   File Location

Notes:
_____
_____
_____
_____
_____
_____

_____
_____
_____
_____
_____
_____
_____
_____

_____
_____
_____
_____
_____
_____
_____
_____
_____

Date	Start: __ / __ / __	Location:	Moon Phase _____
	End: __ / __ / __	GPS:	Brightness: _____ %

Time	Start: _____	Bortle Class: _____	Weather Conditions
	End: _____	Brightness: _____ %	🌡_____ 🎐_____ 💧_____

Object Name:	Constellation:

Catalog Name:	Magnitude Size:

Scope:	Tracker:

Star Alignment: 1)　　　　　　　2)　　　　　　　3)

Camera:	Lens:	Filter/s:	Bahtinov Mask ☐
Exposure:	ISO:	Focal Length:	Aperture:

Lights #_____　　Darks #_____　　BIAS #_____　　Flats #_____　　File Location

Notes:
_____
_____
_____
_____
_____

_____
_____
_____
_____
_____
_____
_____
_____

_____
_____
_____
_____
_____
_____
_____
_____

Date	Start: ___ / ___ / ___	Location:	Moon Phase _____
	End: ___ / ___ / ___	GPS:	Brightness: _____ %
Time	Start: _____	Bortle Class: _____	Weather Conditions
	End: _____	Brightness: _____ %	🌡_____ 🚩_____ 💧_____

Object Name:	Constellation:
Catalog Name:	Magnitude Size:
Scope:	Tracker:

Star Alignment: 1) _____ 2) _____ 3) _____

Camera:	Lens:	Filter/s:	Bahtinov Mask ☐
Exposure:	ISO:	Focal Length:	Aperture:

Lights # _____   Darks # _____   BIAS # _____   Flats # _____ | File Location

Notes:
_____
_____
_____
_____
_____
_____

Date	Start: ___/___/___	Location:	Moon Phase _____
	End: ___/___/___	GPS:	Brightness: _____%
Time	Start: _____	Bortle Class: _____	Weather Conditions
	End: _____	Brightness: _____%	🌡_____ 🚩_____ 💧_____

Object Name:	Constellation:
Catalog Name:	Magnitude Size:
Scope:	Tracker:

Star Alignment: 1)   2)   3)

Camera:	Lens:	Filter/s:	Bahtinov Mask ☐
Exposure:	ISO:	Focal Length:	Aperture:

Lights #_____   Darks #_____   BIAS #_____   Flats #_____   File Location

Notes:
_____
_____
_____
_____
_____
_____

Date	Start: / /	Location:	Moon Phase
	End: / /	GPS:	Brightness: %
Time	Start:	Bortle Class:	Weather Conditions
	End:	Brightness: %	🌡_____ 🚩_____ 💧_____

Object Name:	Constellation:
Catalog Name:	Magnitude Size:
Scope:	Tracker:

Star Alignment: 1)　　　　　　2)　　　　　　3)

Camera:	Lens:	Filter/s:	Bahtinov Mask ☐
Exposure:	ISO:	Focal Length:	Aperture:

Lights #_____　　Darks #_____　　BIAS #_____　　Flats #_____ | File Location

Notes:

_____
_____
_____
_____
_____
_____

_____
_____
_____
_____
_____
_____
_____
_____

_____
_____
_____
_____
_____
_____
_____
_____
_____

Date Start: ___/___/___ End: ___/___/___	Location: GPS:	Moon Phase _____ Brightness: _____ %	

Time Start: _____ End: _____

Bortle Class: _____
Brightness: _____ %

Weather Conditions
🌡 _____ 🚩 _____ 💧 _____

Object Name: _____ Constellation: _____

Catalog Name: _____ Magnitude Size: _____

Scope: _____ Tracker: _____

Star Alignment: 1) _____ 2) _____ 3) _____

Camera:	Lens:	Filter/s:	Bahtinov Mask ☐
Exposure:	ISO:	Focal Length:	Aperture:

Lights #_____ Darks #_____ BIAS #_____ Flats #_____ File Location

Notes:
_____
_____
_____
_____
_____

Date Start: ___ / ___ / ___ End: ___ / ___ / ___	Location: GPS:	Moon Phase _____ Brightness: _____ %	

Time Start: _____ End: _____	Bortle Class: _____ Brightness: _____%	Weather Conditions 🌡️_____ 🚩_____ 💧_____

Object Name: _____ Constellation: _____

Catalog Name: _____ Magnitude Size: _____

Scope: _____ Tracker: _____

Star Alignment: 1) _____ 2) _____ 3) _____

Camera:	Lens:	Filter/s:	Bahtinov Mask ☐
Exposure:	ISO:	Focal Length:	Aperture:

Lights #_____ Darks #_____ BIAS #_____ Flats #_____ File Location

Notes: _____
_____
_____
_____
_____
_____

Date	Start: ___ / ___ / ___	Location:	Moon Phase _____
	End: ___ / ___ / ___	GPS:	Brightness: _____ %

Time	Start: _____	Bortle Class: _____	Weather Conditions
	End: _____	Brightness: _____ %	🌡_____ 🚩_____ 💧_____

Object Name:	Constellation:

Catalog Name:	Magnitude Size:

Scope:	Tracker:

Star Alignment: 1)   2)   3)

Camera:	Lens:	Filter/s:	Bahtinov Mask ☐
Exposure:	ISO:	Focal Length:	Aperture:

Lights #_____   Darks #_____   BIAS #_____   Flats #_____ | File Location

Notes:
_____
_____
_____
_____
_____

Date Start: ___/___/___ End: ___/___/___	Location: GPS:	Moon Phase _____ Brightness: _____ %	
Time Start: _____ End: _____	Bortle Class: _____ Brightness: _____%	Weather Conditions 🌡_____ 🚩_____ 💧_____	

Object Name:

Constellation:

Catalog Name:

Magnitude Size:

Scope:

Tracker:

Star Alignment: 1)   2)   3)

Camera:	Lens:	Filter/s:	Bahtinov Mask ☐
Exposure:	ISO:	Focal Length:	Aperture:

Lights #_____   Darks #_____   BIAS #_____   Flats #_____   File Location

Notes:
_____
_____
_____
_____
_____

Date Start: ___ / ___ / ___	Location:	Moon Phase _____		
Date End: ___ / ___ / ___	GPS:	Brightness: _____ %		
Time Start: _____	Bortle Class: _____	Weather Conditions		
Time End: _____	Brightness: _____ %	🌡_____ 🚩_____ 💧_____		

Object Name:	Constellation:
Catalog Name:	Magnitude Size:
Scope:	Tracker:

Star Alignment: 1)   2)   3)

Camera:	Lens:	Filter/s:	Bahtinov Mask ☐
Exposure:	ISO:	Focal Length:	Aperture:

Lights #_____   Darks #_____   BIAS #_____   Flats #_____ | File Location

Notes:
_____
_____
_____
_____
_____
_____

Date	Start: ___/___/___	Location:	Moon Phase _____
	End: ___/___/___	GPS:	Brightness: _____ %
Time	Start: _____	Bortle Class: _____	Weather Conditions
	End: _____	Brightness: _____ %	🌡_____ 🚩_____ 💧_____

Object Name:	Constellation:
Catalog Name:	Magnitude Size:
Scope:	Tracker:

Star Alignment: 1)   2)   3)

Camera:	Lens:	Filter/s:	Bahtinov Mask ☐
Exposure:	ISO:	Focal Length:	Aperture:

Lights #_____   Darks #_____   BIAS #_____   Flats #_____ | File Location

Notes:
_____
_____
_____
_____
_____

Date Start: ___ / ___ / ___	Location:	Moon Phase _____	
Date End: ___ / ___ / ___	GPS:	Brightness: _____ %	
Time Start: _____	Bortle Class: _____	Weather Conditions	
Time End: _____	Brightness: _____ %	🌡_____ 🚩_____ 💧_____	

Object Name:	Constellation:
Catalog Name:	Magnitude Size:
Scope:	Tracker:

Star Alignment: 1)   2)   3)

Camera:	Lens:	Filter/s:	Bahtinov Mask ☐
Exposure:	ISO:	Focal Length:	Aperture:

Lights # _____   Darks # _____   BIAS # _____   Flats # _____   File Location

Notes:
_____
_____
_____
_____
_____
_____

_____
_____
_____
_____
_____
_____
_____
_____

_____
_____
_____
_____
_____
_____
_____
_____

Date	Start: ___/___/___	Location:	Moon Phase _____
	End: ___/___/___	GPS:	Brightness: _____ %
Time	Start: _____	Bortle Class: _____	Weather Conditions
	End: _____	Brightness: _____ %	🌡️_____ 🚩_____ 💧_____

Object Name:	Constellation:
Catalog Name:	Magnitude Size:
Scope:	Tracker:

Star Alignment: 1)   2)   3)

| Camera: | Lens: | Filter/s: | Bahtinov Mask ☐ |
| Exposure: | ISO: | Focal Length: | Aperture: |

Lights #_____   Darks #_____   BIAS #_____   Flats #_____   File Location

Notes:
_____
_____
_____
_____
_____
_____

Date	Start: ___/___/___	Location:	Moon Phase _____
	End: ___/___/___	GPS:	Brightness: _____ %
Time	Start: _____	Bortle Class: _____	Weather Conditions
	End: _____	Brightness: _____ %	🌡_____ 🚩_____ 💧_____

Object Name:	Constellation:
Catalog Name:	Magnitude Size:
Scope:	Tracker:

Star Alignment: 1) _____ 2) _____ 3) _____

Camera:	Lens:	Filter/s:	Bahtinov Mask ☐
Exposure:	ISO:	Focal Length:	Aperture:
Lights #_____ Darks #_____ BIAS #_____ Flats #_____			File Location

Notes:
_____
_____
_____
_____
_____

_____
_____
_____
_____
_____
_____
_____
_____

_____
_____
_____
_____
_____
_____
_____
_____
_____

Date	Start: / /	Location:	Moon Phase
	End: / /	GPS:	Brightness: %
Time	Start:	Bortle Class:	Weather Conditions
	End:	Brightness: %	🌡 ___ 🚩 ___ 💧 ___

Object Name:	Constellation:
Catalog Name:	Magnitude Size:
Scope:	Tracker:

Star Alignment: 1) 2) 3)

Camera:	Lens:	Filter/s:	Bahtinov Mask ☐
Exposure:	ISO:	Focal Length:	Aperture:

Lights # _____ Darks # _____ BIAS # _____ Flats # _____ | File Location

Notes:
_____
_____
_____
_____
_____
_____

Date	Start: ___/___/___	Location:	Moon Phase _____
	End: ___/___/___	GPS:	Brightness: _____%

Time	Start: _____	Bortle Class: _____	Weather Conditions
	End: _____	Brightness: _____%	🌡_____ 🚩_____ 💧_____

Object Name:	Constellation:

Catalog Name:	Magnitude Size:

Scope:	Tracker:

Star Alignment: 1)   2)   3)

Camera:	Lens:	Filter/s:	Bahtinov Mask ☐
Exposure:	ISO:	Focal Length:	Aperture:

Lights #_____	Darks #_____	BIAS #_____	Flats #_____	File Location

Notes:
_____
_____
_____
_____
_____
_____

Date Start: ___ / ___ / ___ End: ___ / ___ / ___	Location: GPS:	Moon Phase _____ Brightness: _____ %	
Time Start: _____ End: _____	Bortle Class: _____ Brightness: _____ %	Weather Conditions 🌡 _____ 🚩 _____ 💧 _____	

Object Name: _____ Constellation: _____

Catalog Name: _____ Magnitude Size: _____

Scope: _____ Tracker: _____

Star Alignment: 1) _____ 2) _____ 3) _____

Camera:	Lens:	Filter/s:	Bahtinov Mask ☐
Exposure:	ISO:	Focal Length:	Aperture:

Lights # _____   Darks # _____   BIAS # _____   Flats # _____   File Location

Notes:
_____
_____
_____
_____
_____
_____
_____

| Date | Start: ___ / ___ / ___ | Location: | Moon Phase _____ |
| | End: ___ / ___ / ___ | GPS: | Brightness: _____ % |

| Time | Start: _____ | Bortle Class: _____ | Weather Conditions |
| | End: _____ | Brightness: _____ % | 🌡_____ 🚩_____ 💧_____ |

Object Name: | Constellation:

Catalog Name: | Magnitude Size:

Scope: | Tracker:

Star Alignment: 1)   2)   3)

| Camera: | Lens: | Filter/s: | Bahtinov Mask ☐ |
| Exposure: | ISO: | Focal Length: | Aperture: |

Lights #_____   Darks #_____   BIAS #_____   Flats #_____ | File Location

Notes:
_____
_____
_____
_____
_____

_____
_____
_____
_____
_____
_____
_____
_____

_____
_____
_____
_____
_____
_____
_____
_____
_____
_____

Date	Start: ___/___/___	Location:	Moon Phase _____
	End: ___/___/___	GPS:	Brightness: _____ %
Time	Start: _____	Bortle Class: _____	Weather Conditions
	End: _____	Brightness: _____ %	🌡_____ 🎏_____ 💧_____

Object Name:	Constellation:
Catalog Name:	Magnitude Size:
Scope:	Tracker:

Star Alignment: 1)   2)   3)

Camera:	Lens:	Filter/s:	Bahtinov Mask ☐
Exposure:	ISO:	Focal Length:	Aperture:

Lights #_____   Darks #_____   BIAS #_____   Flats #_____   File Location

Notes:
_____
_____
_____
_____
_____
_____

Date	Start: ___/___/___	Location:	Moon Phase _____
	End: ___/___/___	GPS:	Brightness: _____ %
Time	Start: _____	Bortle Class: _____	Weather Conditions
	End: _____	Brightness: _____ %	🌡_____ 🚩_____ 💧_____

Object Name:	Constellation:
Catalog Name:	Magnitude Size:
Scope:	Tracker:

Star Alignment: 1)   2)   3)

Camera:	Lens:	Filter/s:	Bahtinov Mask ☐
Exposure:	ISO:	Focal Length:	Aperture:

| Lights # _____ | Darks # _____ | BIAS # _____ | Flats # _____ | File Location |

Notes:
_____
_____
_____
_____
_____

_____
_____
_____
_____
_____
_____
_____
_____
_____

_____
_____
_____
_____
_____
_____
_____
_____
_____
_____

Date Start: ___ / ___ / ___ End: ___ / ___ / ___	Location: GPS:	Moon Phase _____ Brightness: _____ %	
Time Start: _____ End: _____	Bortle Class: _____ Brightness: _____ %	Weather Conditions 🌡_____ 🚩_____ 💧_____	

Object Name:	Constellation:
Catalog Name:	Magnitude Size:
Scope:	Tracker:

Star Alignment: 1)   2)   3)

Camera:	Lens:	Filter/s:	Bahtinov Mask ☐
Exposure:	ISO:	Focal Length:	Aperture:

Lights #_____	Darks #_____	BIAS #_____	Flats #_____	File Location

Notes:
_____
_____
_____
_____
_____
_____

	Start: ___ / ___ / ___	Location:	Moon Phase _____
Date	End: ___ / ___ / ___	GPS:	Brightness: _____ %
Time	Start: _____	Bortle Class: _____	Weather Conditions
	End: _____	Brightness: _____ %	🌡_____ 🚩_____ 💧_____

Object Name: _____ Constellation: _____

Catalog Name: _____ Magnitude Size: _____

Scope: _____ Tracker: _____

Star Alignment: 1)   2)   3)

Camera:	Lens:	Filter/s:	Bahtinov Mask ☐
Exposure:	ISO:	Focal Length:	Aperture:

Lights #_____   Darks #_____   BIAS #_____   Flats #_____   File Location

Notes:
_____
_____
_____
_____
_____
_____
_____

_____
_____
_____
_____
_____
_____
_____
_____

_____
_____
_____
_____
_____
_____
_____

Date	Start: / /	Location:	Moon Phase _____
	End: / /	GPS:	Brightness: _____ %
Time	Start: _____	Bortle Class: _____	Weather Conditions
	End: _____	Brightness: _____%	🌡_____ 🚩_____ 💧_____

Object Name: | Constellation:

Catalog Name: | Magnitude Size:

Scope: | Tracker:

Star Alignment: 1)   2)   3)

| Camera: | Lens: | Filter/s: | Bahtinov Mask ☐ |
| Exposure: | ISO: | Focal Length: | Aperture: |

Lights #_____   Darks #_____   BIAS #_____   Flats #_____ | File Location

Notes:
_____
_____
_____
_____
_____
_____

_____
_____
_____
_____
_____
_____
_____
_____

_____
_____
_____
_____
_____
_____
_____

Date	Start: ___/___/___	Location:	Moon Phase _____
	End: ___/___/___	GPS:	Brightness: _____ %

Time	Start: _____	Bortle Class: _____	Weather Conditions
	End: _____	Brightness: _____%	🌡_____ 🚩_____ 💧_____

Object Name:	Constellation:
Catalog Name:	Magnitude Size:
Scope:	Tracker:

Star Alignment: 1) _____ 2) _____ 3) _____

Camera:	Lens:	Filter/s:	Bahtinov Mask ☐
Exposure:	ISO:	Focal Length:	Aperture:

Lights # _____   Darks # _____   BIAS # _____   Flats # _____   File Location

Notes:
_____
_____
_____
_____
_____
_____

	Start: ___/___/___	Location:	Moon Phase _____
Date	End: ___/___/___	GPS:	Brightness: _____ %
Time	Start: _____	Bortle Class: _____	Weather Conditions
	End: _____	Brightness: _____ %	🌡_____ 🚩_____ 💧_____

Object Name:	Constellation:
Catalog Name:	Magnitude Size:
Scope:	Tracker:

Star Alignment: 1)   2)   3)

Camera:	Lens:	Filter/s:	Bahtinov Mask ☐
Exposure:	ISO:	Focal Length:	Aperture:

Lights #_____   Darks #_____   BIAS #_____   Flats #_____ | File Location

Notes:
_____
_____
_____
_____
_____
_____

	Start: ___ / ___ / ___	Location:	Moon Phase _____
Date	End: ___ / ___ / ___	GPS:	Brightness: _____ %
Time	Start: _____	Bortle Class: _____	**Weather Conditions**
	End: _____	Brightness: _____ %	🌡 _____ 🚩 _____ 💧 _____

Object Name: _____ Constellation: _____

Catalog Name: _____ Magnitude Size: _____

Scope: _____ Tracker: _____

Star Alignment: 1) _____ 2) _____ 3) _____

Camera:	Lens:	Filter/s:	Bahtinov Mask ☐
Exposure:	ISO:	Focal Length:	Aperture:

Lights #_____ Darks #_____ BIAS #_____ Flats #_____ | File Location

Notes:
_____
_____
_____
_____
_____

Date	Start: ___/___/___	Location:	Moon Phase ___
	End: ___/___/___	GPS:	Brightness: ___%
Time	Start: ___	Bortle Class: ___	Weather Conditions
	End: ___	Brightness: ___%	🌡___ 🚩___ 💧___

Object Name:	Constellation:
Catalog Name:	Magnitude Size:
Scope:	Tracker:

Star Alignment: 1)   2)   3)

| Camera: | Lens: | Filter/s: | Bahtinov Mask ☐ |
| Exposure: | ISO: | Focal Length: | Aperture: |

Lights #_____   Darks #_____   BIAS #_____   Flats #_____   File Location

Notes:
_____
_____
_____
_____
_____
_____

Date	Start: ___/___/___	Location:	Moon Phase _____
	End: ___/___/___	GPS:	Brightness: _____ %
Time	Start: _____	Bortle Class: _____	**Weather Conditions**
	End: _____	Brightness: _____%	🌡_____ 🎏_____ 💧_____

Object Name:	Constellation:
Catalog Name:	Magnitude Size:
Scope:	Tracker:

Star Alignment: 1) _____ 2) _____ 3) _____

| Camera: | Lens: | Filter/s: | Bahtinov Mask ☐ |
| Exposure: | ISO: | Focal Length: | Aperture: |

Lights #_____   Darks #_____   BIAS #_____   Flats #_____   File Location

Notes:
_____
_____
_____
_____
_____

Date	Start: ___ / ___ / ___	Location:	Moon Phase _____
	End: ___ / ___ / ___	GPS:	Brightness: _____ %
Time	Start: _____	Bortle Class: _____	Weather Conditions
	End: _____	Brightness: _____ %	🌡_____ 🚩_____ 💧_____

Object Name:	Constellation:
Catalog Name:	Magnitude Size:
Scope:	Tracker:

Star Alignment: 1)   2)   3)

Camera:	Lens:	Filter/s:	Bahtinov Mask ☐
Exposure:	ISO:	Focal Length:	Aperture:

Lights #_____   Darks #_____   BIAS #_____   Flats #_____   File Location

Notes:
_____
_____
_____
_____
_____
_____

_____
_____
_____
_____
_____
_____
_____
_____
_____

_____
_____
_____
_____
_____
_____
_____
_____
_____
_____

Date	Start: ___/___/___	Location:	Moon Phase _____
	End: ___/___/___	GPS:	Brightness: _____ %

Time	Start: _____	Bortle Class: _____	Weather Conditions
	End: _____	Brightness: _____ %	🌡_____ 🎏_____ 💧_____

Object Name: _____ Constellation: _____

Catalog Name: _____ Magnitude Size: _____

Scope: _____ Tracker: _____

Star Alignment: 1) _____ 2) _____ 3) _____

Camera:	Lens:	Filter/s:	Bahtinov Mask ☐
Exposure:	ISO:	Focal Length:	Aperture:

Lights #_____   Darks #_____   BIAS #_____   Flats #_____   File Location

Notes:
_____
_____
_____
_____
_____
_____

Date	Start: ___/___/___	Location:	Moon Phase _____
	End: ___/___/___	GPS:	Brightness: _____ %
Time	Start: _____	Bortle Class: _____	Weather Conditions
	End: _____	Brightness: _____ %	🌡_____ 🚩_____ 💧_____

Object Name:	Constellation:
Catalog Name:	Magnitude Size:
Scope:	Tracker:

Star Alignment: 1)   2)   3)

Camera:	Lens:	Filter/s:	Bahtinov Mask ☐
Exposure:	ISO:	Focal Length:	Aperture:

Lights #_____   Darks #_____   BIAS #_____   Flats #_____   File Location

Notes:
_____
_____
_____
_____
_____
_____

_____
_____
_____
_____
_____
_____
_____
_____

_____
_____
_____
_____
_____
_____
_____
_____
_____

Date	Start: ___/___/___	Location:	Moon Phase _____
	End: ___/___/___	GPS:	Brightness: _____ %
Time	Start: _____	Bortle Class: _____	Weather Conditions
	End: _____	Brightness: _____ %	🌡_____ 💨_____ 💧_____

Object Name:	Constellation:
Catalog Name:	Magnitude Size:
Scope:	Tracker:

Star Alignment: 1)　　　　　　　2)　　　　　　　3)

Camera:	Lens:	Filter/s:	Bahtinov Mask ☐
Exposure:	ISO:	Focal Length:	Aperture:

Lights #_____　　Darks #_____　　BIAS #_____　　Flats #_____ | File Location

Notes:
_____
_____
_____
_____
_____

_____
_____
_____
_____
_____
_____
_____
_____
_____

_____
_____
_____
_____
_____
_____
_____
_____
_____

Date	Start: ___/___/___	Location:	Moon Phase _____
	End: ___/___/___	GPS:	Brightness: _____ %
Time	Start: _____	Bortle Class: _____	Weather Conditions
	End: _____	Brightness: _____%	🌡_____ 🚩_____ 💧_____

Object Name:	Constellation:
Catalog Name:	Magnitude Size:
Scope:	Tracker:

Star Alignment: 1)   2)   3)

Camera:	Lens:	Filter/s:	Bahtinov Mask ☐
Exposure:	ISO:	Focal Length:	Aperture:

Lights #_____   Darks #_____   BIAS #_____   Flats #_____ | File Location

Notes:
_____
_____
_____
_____
_____
_____

Date	Start: ___ / ___ / ___	Location:	Moon Phase _____
	End: ___ / ___ / ___	GPS:	Brightness: _____ %
Time	Start: _____	Bortle Class: _____	Weather Conditions
	End: _____	Brightness: _____ %	🌡 _____ ⚑ _____ 💧 _____

Object Name:	Constellation:
Catalog Name:	Magnitude Size:
Scope:	Tracker:

Star Alignment: 1)　　　　　　　　2)　　　　　　　　3)

Camera:	Lens:	Filter/s:	Bahtinov Mask ☐
Exposure:	ISO:	Focal Length:	Aperture:

Lights #_____　　Darks #_____　　BIAS #_____　　Flats #_____　　File Location

Notes:
_____
_____
_____
_____
_____
_____

Date	Start: ___ / ___ / ___	Location:	Moon Phase _____
	End: ___ / ___ / ___	GPS:	Brightness: _____ %
Time	Start: _____	Bortle Class: _____	Weather Conditions
	End: _____	Brightness: _____ %	🌡_____ 🚩_____ 💧_____

Object Name:	Constellation:
Catalog Name:	Magnitude Size:
Scope:	Tracker:

Star Alignment: 1)   2)   3)

| Camera: | Lens: | Filter/s: | Bahtinov Mask ☐ |
| Exposure: | ISO: | Focal Length: | Aperture: |

Lights # _____   Darks # _____   BIAS # _____   Flats # _____   File Location

Notes:
_____
_____
_____
_____
_____
_____

Date	Start: ___/___/___	Location:	Moon Phase _____
	End: ___/___/___	GPS:	Brightness: _____ %
Time	Start: _____	Bortle Class: _____	Weather Conditions
	End: _____	Brightness: _____ %	🌡_____ 🚩_____ 💧_____

Object Name: _____ Constellation: _____

Catalog Name: _____ Magnitude Size: _____

Scope: _____ Tracker: _____

Star Alignment: 1) _____ 2) _____ 3) _____

Camera:	Lens:	Filter/s:	Bahtinov Mask ☐
Exposure:	ISO:	Focal Length:	Aperture:

Lights #_____   Darks #_____   BIAS #_____   Flats #_____ | File Location

Notes: _____
_____
_____
_____
_____
_____

	Start: ___/___/___	Location:	Moon Phase _____
Date	End: ___/___/___	GPS:	Brightness: _____ %
Time	Start: _____	Bortle Class: _____	Weather Conditions
	End: _____	Brightness: _____ %	🌡_____ 🚩_____ 💧_____

Object Name:	Constellation:
Catalog Name:	Magnitude Size:
Scope:	Tracker:

Star Alignment: 1)   2)   3)

| Camera: | Lens: | Filter/s: | Bahtinov Mask ☐ |
| Exposure: | ISO: | Focal Length: | Aperture: |

Lights #_____   Darks #_____   BIAS #_____   Flats #_____   File Location

Notes:
_____
_____
_____
_____
_____
_____

_____
_____
_____
_____
_____
_____
_____
_____

_____
_____
_____
_____
_____
_____
_____
_____
_____

	Start: ___/___/___	Location:	Moon Phase _____
Date	End: ___/___/___	GPS:	Brightness: _____ %
Time	Start: _____	Bortle Class: _____	Weather Conditions
	End: _____	Brightness: _____ %	🌡_____ 💨_____ 💧_____

Object Name:	Constellation:
Catalog Name:	Magnitude Size:
Scope:	Tracker:

Star Alignment: 1)　　　　　　　2)　　　　　　　3)

Camera:	Lens:	Filter/s:	Bahtinov Mask ☐
Exposure:	ISO:	Focal Length:	Aperture:

Lights #_____　　Darks #_____　　BIAS #_____　　Flats #_____　　File Location

Notes:
_____
_____
_____
_____
_____
_____

Date	Start: ___ / ___ / ___	Location:	Moon Phase _____
	End: ___ / ___ / ___	GPS:	Brightness: _____ %
Time	Start: _____	Bortle Class: _____	Weather Conditions
	End: _____	Brightness: _____ %	🌡_____ 🚩_____ 💧_____

Object Name:	Constellation:
Catalog Name:	Magnitude Size:
Scope:	Tracker:

Star Alignment: 1)   2)   3)

| Camera: | Lens: | Filter/s: | Bahtinov Mask ☐ |
| Exposure: | ISO: | Focal Length: | Aperture: |

Lights #_____   Darks #_____   BIAS #_____   Flats #_____   File Location

Notes:
_____
_____
_____
_____
_____
_____

Date	Start: ___/___/___	Location:	Moon Phase _____
	End: ___/___/___	GPS:	Brightness: _____ %
Time	Start: _____	Bortle Class: _____	Weather Conditions
	End: _____	Brightness: _____%	🌡_____ 🚩_____ 💧_____

Object Name:	Constellation:
Catalog Name:	Magnitude Size:
Scope:	Tracker:

Star Alignment: 1)   2)   3)

Camera:	Lens:	Filter/s:	Bahtinov Mask ☐
Exposure:	ISO:	Focal Length:	Aperture:

Lights #_____   Darks #_____   BIAS #_____   Flats #_____ | File Location

Notes:
_____
_____
_____
_____
_____

_____
_____
_____
_____
_____
_____
_____
_____

_____
_____
_____
_____
_____
_____
_____
_____

Date	Start: ___/___/___	Location:	Moon Phase _____
	End: ___/___/___	GPS:	Brightness: _____ %
Time	Start: _____	Bortle Class: _____	Weather Conditions
	End: _____	Brightness: _____ %	🌡_____ 🚩_____ 💧_____

Object Name:	Constellation:
Catalog Name:	Magnitude Size:
Scope:	Tracker:

Star Alignment: 1)   2)   3)

| Camera: | Lens: | Filter/s: | Bahtinov Mask ☐ |
| Exposure: | ISO: | Focal Length: | Aperture: |

Lights # _____   Darks # _____   BIAS # _____   Flats # _____   File Location

Notes:
_____
_____
_____
_____
_____
_____

Date	Start: ___ / ___ / ___	Location:	Moon Phase _____
	End: ___ / ___ / ___	GPS:	Brightness: _____ %
Time	Start: _____	Bortle Class: _____	Weather Conditions
	End: _____	Brightness: _____ %	🌡_____ 🚩_____ 💧_____

Object Name:	Constellation:
Catalog Name:	Magnitude Size:
Scope:	Tracker:

Star Alignment: 1)　　　　　　2)　　　　　　3)

Camera:	Lens:	Filter/s:	Bahtinov Mask ☐
Exposure:	ISO:	Focal Length:	Aperture:

Lights #_____　　Darks #_____　　BIAS #_____　　Flats #_____　　File Location

Notes:
_____
_____
_____
_____
_____
_____

Date	Start: ___/___/___	Location:	Moon Phase _____
	End: ___/___/___	GPS:	Brightness: _____ %
Time	Start: _____	Bortle Class: _____	Weather Conditions
	End: _____	Brightness: _____ %	🌡_____ 🚩_____ 💧_____

Object Name:	Constellation:
Catalog Name:	Magnitude Size:
Scope:	Tracker:

Star Alignment: 1)   2)   3)

Camera:	Lens:	Filter/s:	Bahtinov Mask ☐
Exposure:	ISO:	Focal Length:	Aperture:

Lights #_____   Darks #_____   BIAS #_____   Flats #_____   File Location

Notes:
_____
_____
_____
_____
_____
_____

_____
_____
_____
_____
_____
_____
_____
_____

_____
_____
_____
_____
_____
_____
_____
_____
_____

EXPERIMENTAL FIELD SETTINGS

EQUIPMENT	RESULTS

EXPERIMENTAL FIELD SETTINGS

EQUIPMENT	RESULTS

NOTES

NOTES

Fellow Companion Salutations

Made in the USA
Columbia, SC
02 November 2024

45540010R00070